The Splitting of the Moon: A Chronicle of Celestial Ruin and Human Resilience

The Splitting of the Moon

A.N.F. Simões

Published by OLYSSIPO, 2024.

While every precaution has been taken in the preparation of this book, the publisher assumes no responsibility for errors or omissions, or for damages resulting from the use of the information contained herein.

THE SPLITTING OF THE MOON

First edition. November 22, 2024.

Copyright © 2024 A.N.F. Simões.

ISBN: 979-8230376071

Written by A.N.F. Simões.

Also by A.N.F. Simões

In the Shadow of the Kurgan: A Compendium of the
Indigenous People of Europe
The Cornfield
The Secret Rout
The Time Weaver's Tapestry
The New Republic of Manhood: Building a Better Future in a
World Gone Wrong
The Splitting of the Moon

Table of Contents

The Splitting of the Moon 1

Preface 3

Chapter 1: The Shadow of 'Mohammed' 7

Chapter 2: Echoes of Celestial Upheaval 11

Chapter 3: Celestial Omens and Human Folly 15

Chapter 4: Vestiges of a Vanished Orb: Speculations on a Pre-Lunar Earth 21

Chapter 5: The Moon's Eternal Vigil 25

Chapter 6: Reflections on a Celestial Cataclysm: Humanity's Future Amongst the Stars 33

Chapter 7: A Moonless World: Imagining the Unthinkable ... 41

Chapter 8: Homo Lunaris: Humanity in a Moonless Future ... 49

Synopsis

"The Splitting of the Moon" explores the hypothetical scenario of a large meteor impacting the Moon and its profound consequences for Earth and human civilization. This work of speculative non-fiction blends scientific rigor with philosophical reflection, examining the astronomical, environmental, social, and cultural ramifications of a world suddenly deprived of its celestial companion. The narrative begins by tracing humanity's historical fascination with the cosmos, exploring the myths, legends, and religious beliefs that have shaped our understanding of the Moon and its place in the universe. It then delves into the scientific realities of a lunar impact, analyzing the potential geological, atmospheric, and tidal effects of such a cataclysmic event. The hypothetical impact serves as a catalyst for exploring the challenges and opportunities facing a world grappling with a fundamentally altered celestial landscape. The book examines the potential for technological innovation, international cooperation, and cultural adaptation as humanity confronts the loss of its familiar lunar companion. The narrative also delves into the psychological and philosophical implications of a Moonless world, exploring the potential for a shift in human consciousness, a renewed appreciation for the fragility of our planet, and a deeper understanding of our place in the cosmos. "The Splitting of the Moon" is a work of both warning and hope. It serves as a reminder of the dynamic nature of our cosmic environment and the potential for catastrophic events beyond our control. But it also highlights the resilience and adaptability of human civilization, our capacity to confront the unknown, and our enduring quest for knowledge and meaning in a universe full of wonder and uncertainty. "The Splitting of the Moon" is a thought-provoking exploration of a hypothetical scenario that challenges our assumptions about the stability of our cosmic environment and invites us to contemplate the future of humanity in a universe full of wonder and uncertainty.

Preface

The Moon, that luminous orb traversing the celestial canvas, has been an object of wonder and fascination for humanity since time immemorial. Its silvery light has illuminated our nights, guided our wanderings, and inspired our dreams. From the lunar deities of antiquity to the modern marvels of lunar exploration, the Moon has held a mirror to our hopes, fears, and aspirations. Across cultures and epochs, we have woven intricate tapestries of myth and legend around the Moon. In the East, the Moon was often personified as a deity, imbued with divine powers and mystical attributes. In Hindu cosmology, Chandra, the lunar god, rode his chariot across the night sky, his brilliance waxing and waning with the cycle of life and death. The Chinese revered Chang'e, the Moon goddess, who resided in a palace of jade, her ethereal beauty a symbol of immortality.

Western civilizations also crafted rich narratives around the Moon. The Greeks worshipped Selene, the lunar goddess, who drove her moon-chariot across the heavens, her silvery light casting an enchanting spell upon the world. The Romans honored Luna, her counterpart, as a protector of travelers and a guardian of the night. In Norse mythology, Máni, the Moon god, was pursued across the sky by a monstrous wolf, their eternal chase symbolizing the cyclical nature of time and the ever-present threat of chaos.

Beyond mythology, celestial events have long been interpreted as omens, harbingers of change, and warnings of impending doom. Comets, with their fiery tails streaking across the heavens, were often seen as messengers of ill fortune, their appearance heralding war, famine, or the fall of empires. Eclipses, those moments of celestial drama when the Sun or Moon is obscured, were viewed with trepidation, their sudden darkness casting a pall over the world. Even the seemingly predictable movements of the stars were scrutinized for hidden meanings, astrologers and soothsayers claiming to decipher the celestial script and predict the fate of individuals and nations.

The Moon, with its cyclical phases and gravitational influence on Earth's tides, held a particular sway over human imagination, inspiring countless myths and legends. In the Islamic tradition, the Moon holds a particularly profound significance. It is enshrined in the celestial tapestry of the Quran, its crescent a symbol of faith and guidance. And it is the protagonist of a miraculous event, a testament to the Prophet Mohammed's divine connection: the splitting of the Moon. This extraordinary act, defying the laws of nature, has resonated through centuries, inspiring awe and reaffirming belief.

This work, however, ventures beyond the realm of the miraculous and into the domain of the plausible, yet no less awe-inspiring. We embark on a journey of imagination, guided by the spirit of historical inquiry and scientific curiosity, to explore a hypothetical event of cataclysmic proportions: the splitting of the Moon, not by divine intervention, but by the brute force of a celestial wanderer.

Imagine a future where humanity, in its relentless pursuit of progress, has reached for the stars, establishing a foothold on the lunar surface. Imagine a world interconnected by a delicate web of technology, reliant on the celestial ballet of Earth and Moon. Then, imagine this delicate balance shattered by a rogue meteor, a cosmic interloper bearing the ironic name of "Mohammed."

What follows is a chronicle of celestial ruin and human resilience. We shall witness the terrifying grandeur of the impact, the devastation wrought upon our planet, and the profound challenges faced by a civilization grappling with the loss of its celestial companion. We shall explore the scientific, social, and philosophical implications of this event, drawing parallels to historical catastrophes and examining the enduring human capacity for adaptation and innovation.

In the spirit of our intellectual forebears, we aim to capture the grand sweep of history, the interplay of fate and human agency, and the enduring human spirit in the face of overwhelming odds. This is a story of destruction and renewal, a testament to the enduring fascination with the Moon and a reflection on our place in the vast and unpredictable cosmos.

Chapter 1: The Shadow of 'Mohammed'

In the annals of Islamic history, amidst the narratives of revelation and prophecy, there lies an account of a celestial wonder, a testament to the divine power bestowed upon the Prophet Mohammed: the splitting of the Moon. This event, enshrined in the verses of the Quran and recounted in the Hadith, holds profound significance for Muslims, serving as a symbol of faith, a confirmation of prophethood, and a demonstration of Allah's omnipotence.

The setting for this miraculous event was Mina, a valley near Mecca, during the time of the Prophet's mission to spread the message of Islam. As recounted in the Hadith, a group of Meccan pagans, skeptical of Mohammed's claims of prophethood, challenged him to perform a miracle as proof of his divine connection. They demanded a sign, a demonstration of power that would defy the laws of nature and convince them of his authenticity.

In response to this challenge, the Prophet Mohammed, with unwavering faith and divine guidance, raised his hands towards the heavens and invoked Allah's power. And then, before the astonished eyes of the assembled crowd, the Moon, that celestial beacon in the night sky, was split asunder. The two halves of the Moon, separated by a visible gap, stood as a celestial testimony to the Prophet's divine connection and the omnipotence of Allah.

The Quran, the holy book of Islam, alludes to this event in Surah Al-Qamar (The Moon), verses 1-2: "The Hour has drawn near, and the moon has been split asunder. And if they see a sign, they turn away and say, 'This is continuous magic.'" These verses not only confirm the occurrence of the miracle but also highlight the skepticism and denial of those who witnessed it, choosing to attribute it to sorcery rather than divine intervention.

The Hadith, the collected sayings and traditions of the Prophet Mohammed, provide further details of this extraordinary event. According to various narrations, the splitting of the Moon was witnessed by numerous individuals, including companions of the Prophet and even some of his skeptics. Some accounts describe the two halves of the Moon appearing on opposite sides of a mountain, while others mention the appearance of a luminous object between the separated halves.

Despite the eyewitness accounts and the Quranic confirmation, the splitting of the Moon remains a subject of debate and skepticism among some non-Muslim scholars and historians. They question the lack of corroborating evidence from other cultures and the feasibility of such an event from a scientific perspective. However, for Muslims, the splitting of the Moon is an article of faith, a testament to the Prophet's divine connection and a reminder of Allah's power to transcend the laws of nature.

The significance of this miracle extends beyond its historical context. It serves as a powerful symbol of faith, inspiring believers to trust in Allah's omnipotence and the truthfulness of the Prophet's message. It also highlights the limitations of human understanding and the potential for divine intervention in the natural order.

The splitting of the Moon, a celestial wonder witnessed by many, remains a testament to the enduring power of faith and a reminder of the profound connection between the divine and the earthly realms. It is a story that has resonated through centuries, inspiring awe, reaffirming belief, and serving as a beacon of hope for those who seek guidance in the vast and mysterious cosmos.

Chapter 2: Echoes of Celestial Upheaval

In the vast and intricate tapestry of human history, where the threads of myth and reality intertwine, the celestial dome above has served as an endless source of wonder, fear, and inspiration. The hypothetical rending of the Moon, that pearly orb that has illuminated the nocturnal realm since time immemorial, evokes echoes of ancient narratives, whispered tales of divine intervention and cosmic upheaval that have reverberated through the ages.

Across the expanse of human civilizations, from the sun-drenched plains of Mesopotamia to the mist-shrouded mountains of the Andes, the celestial bodies have been imbued with profound significance. The Sun, that radiant beacon of life, and the Moon, that serene guardian of the night, have been worshipped as deities, revered as ancestors, and invoked in countless myths and legends. These celestial narratives, woven into the very fabric of human culture, reflect our enduring fascination with the cosmos and our ceaseless quest to comprehend our place within the grand scheme of the universe

In the annals of Islamic tradition, the alleged splitting of the Moon stands as a testament to the Prophet Mohammed's divine connection and Allah's omnipotence. This miraculous event, enshrined in the sacred verses of the Quran and the venerable Hadith, serves as a beacon of faith, a confirmation of prophethood, and a poignant reminder of the potential fc divine intervention in the natural order. It is a narrative tha has resonated through the centuries, inspiring awe and devotio: among the faithful.

The rich and variegated tapestry of Hindu mythology, with it: pantheon of gods and goddesses and its treasure trove of epic tales, offers a plethora of celestial interventions that resonate with the Islamic account of the Moon's division. The Churning of the Ocean, that primordial struggle between the gods and demons, speaks of a cosmic upheaval that reshaped the very fabric of existence. Lord Krishna's lifting of Mount Govardhan, a feat of divine strength and compassion, demonstrates the power of the divine to protect and preserve. The descent of the Ganges, that sacred river that flows from the heavens to Earth, is another testament to the divine hand that shapes the destiny of humankind. And the eclipses, attributed to the insatiable demon Rahu, serve as reminders of the cosmic forces that govern our lives.

Beyond the realms of Islam and Hinduism, countless other cultures and religions bear witness to the enduring human fascination with celestial intervention. In the myths of ancient Greece, the god Apollo guides his fiery chariot across the sky, bringing light and warmth to the world. In the sagas of the Norsemen, Thor, the mighty thunder god, battles giants and monsters, his hammer Mjolnir striking fear into the hearts of the wicked. And in the oral traditions of indigenous cultures across the globe, celestial bodies are venerated as ancestors or spirits, their movements and appearances imbued with profound spiritual significance.

These narratives, though diverse in their cultural contexts and specific details, share common threads that bind them together. They reflect our innate curiosity about the cosmos, our attempts to grapple with the mysteries of the universe, and our belief in the existence of forces beyond our comprehension. They serve as reminders of our place within the grand tapestry of existence, both humbling us with our insignificance and inspiring us with the potential for greatness.

In an age of scientific enlightenment, where reason and observation reign supreme, it is tempting to dismiss these ancient narratives as mere fables, relics of a bygone era. Yet, their enduring power lies not in their literal truth but in their symbolic resonance. They speak to the deepest yearnings of the human spirit, our longing for meaning, purpose, and connection to something greater than ourselves.

The hypothetical splitting of the Moon, though grounded in the realm of scientific inquiry, taps into this rich vein of mythological and religious symbolism. It is a narrative that speaks to our fascination with the cosmos, our vulnerability to the forces of nature, and our enduring quest for meaning in a universe that often seems indifferent to our fate. It is a story that reminds us of the fragility of our existence and the power of the human spirit to persevere in the face of adversity.

As we stand on the precipice of a new era, one shaped by the hypothetical loss of our celestial companion, we would do well to heed the wisdom embedded in these ancient tales. For they offer not only explanations of the unexplained but also profound insights into the human condition. They remind us that even in the face of cosmic upheaval, the human spirit can endure, adapt, and ultimately find new meaning and purpose in a world forever transformed.

Chapter 3: Celestial Omens and Human Folly

The celestial sphere, that vast expanse studded with a myriad of stars and traversed by wandering planets, has captivated humanity since time immemorial. Ancient civilizations, lacking the sophisticated instruments of modern astronomy, gazed upon this cosmic panorama with a mixture of awe and trepidation. They sought to understand the celestial dance, to decipher its patterns and interpret its meaning, often weaving their observations into intricate tapestries of myth, religion, and cultural narratives.

Comets, those celestial interlopers with their fiery tails streaking across the heavens, were often viewed with a sense of foreboding. Their sudden appearances, defying the predictable rhythms of the stars and planets, were interpreted as divine warnings, portents of war, famine, or the fall of empires. The Babylonians, meticulous observers of the night sky, recorded cometary appearances as far back as the first millennium BC, associating them with earthly upheavals and political turmoil. In ancient Rome, the appearance of a comet was believed to have presaged the assassination of Julius Caesar, while the Great Comet of 1066, immortalized in the Bayeux Tapestry, was seen as an omen of the Norman Conquest of England.

Eclipses, those moments of celestial drama when the Sun or Moon is obscured, were also viewed with apprehension. The sudden darkness, disrupting the natural order of day and night, was often interpreted as a sign of divine displeasure or an impending catastrophe. In ancient China, eclipses were believed to be caused by a celestial dragon devouring the Sun or Moon, requiring elaborate rituals and ceremonies to appease the celestial forces and restore balance to the cosmos. The ancient Greeks, with their philosophical bent, viewed eclipses as disruptions in the cosmic harmony, portending misfortune and societal upheaval.

Even the seemingly predictable movements of the stars were scrutinized for hidden meanings. Astrology, the belief that celestial bodies influence human affairs, flourished in many ancient cultures. Astrologers, claiming to decipher the celestial script, advised rulers, predicted the fate of individuals and nations, and guided decisions on matters of war, peace, and agriculture. While astrology has been largely discredited by modern science, its enduring influence can be seen in our continued fascination with horoscopes and the persistence of astrological beliefs in popular culture.

The Moon, Earth's celestial companion, held a particular sway over human imagination. Its cyclical phases, its gravitational influence on our tides, and its serene beauty have inspired countless myths, legends, and artistic expressions. In many cultures, the Moon was personified as a deity, imbued with divine powers and mystical attributes.

Yet, despite our scientific advancements, a lingering fascination with celestial events persists. The appearance of a comet still evokes a sense of wonder and awe, even if we no longer fear it as a harbinger of doom. Eclipses, though predictable, continue to draw crowds of observers, eager to witness the celestial dance of Sun, Moon, and Earth. And the Moon, that constant companion in our night sky, continues to inspire artists, poets, and dreamers. Its serene beauty, its gravitational pull on our oceans, and its role in our planet's delicate ecosystem serve as a reminder of our interconnectedness with the cosmos.

But what if this celestial harmony were to be disrupted? What if the Moon, that symbol of constancy, were to be shattered by a catastrophic event? This work explores such a hypothetical scenario, examining the potential consequences of a large meteor impacting the Moon and its implications for Earth and human civilization.

Let us embark on a journey of historical inquiry and scientific speculation. We shall delve into the astronomical realities of such an event, analyze its potential effects on our planet's delicate ecosystem, and explore the social, political, and cultural ramifications of a world forever altered by the loss of its celestial companion.

This is not a tale of science fiction, but a work of speculative non-fiction, grounded in scientific principles and historical analysis. It is a chronicle of potential celestial ruin and human resilience, a testament to the enduring human spirit in the face of cosmic challenges.

Chapter 4: Vestiges of a Vanished Orb: Speculations on a Pre-Lunar Earth

Whilst the learned men of our age, armed with the instruments of astronomy and the calculations of celestial mechanics, pronounce with unwavering certainty upon the Moon's enduring presence as Earth's satellite, there linger, amidst the dusty tomes of antiquity and the whispers of forgotten lore, tantalizing hints of a time when our planet traversed the celestial void unaccompanied by its silvery companion.

Indeed, certain venerable traditions, passed down through generations of bards and chroniclers, speak of a primordial epoch when the night sky was devoid of the Moon's familiar glow, a time when the stars reigned supreme in the unchallenged dominion of darkness. These accounts, though often shrouded in the mists of mythology and embellished by the poetic license of antiquity, nonetheless beckon the curious mind to contemplate the possibility of a pre-lunar Earth, a world where the tides ebbed and flowed not to the rhythm of the Moon's celestial dance, but to the more subtle symphony of the Sun's distant embrace.

Amongst the ancient Hellenes, those progenitors of Western philosophy and art, we find whispers of a time before the Moon graced the heavens. The Arcadians, a people dwelling in the rugged mountains of the Peloponnese, claimed with ancestral pride to be *proselenoi,* "before the Moon," their lineage stretching back to an era when the Earth knew only the Sun's diurnal embrace and the stars' nocturnal reign. Aristotle himself, that paragon of intellectual curiosity, alluded to a time when the Moon was absent from the celestial sphere, a notion echoed by later scholars such as Apollonius of Rhodes and Plutarch, whose writings hint at a primordial world bathed in an uninterrupted tapestry of starlight.

Nor are such accounts confined to the shores of the Mediterranean. Across vast oceans and continents, amidst the oral traditions of diverse cultures, we find echoes of a Moonless past. Certain Native American tribes, those custodians of ancient wisdom passed down through generations of storytellers, speak of a time when the night sky was devoid of the Moon's silvery disc, a time when the stars shone with unparalleled brilliance, their light undimmed by the lunar glow. Even the geological record, that stony chronicle of Earth's deep history, has been scrutinized for clues to a pre-lunar past. Some have argued that the apparent absence of lunar tides in certain ancient sedimentary formations suggests a time when the Moon's gravitational influence was absent, a tantalizing hint of a world where the oceans ebbed and flowed to a different rhythm. Yet, the pronouncements of modern science, with its arsenal of telescopes, spectrometers, and computer simulations, stand in stark contrast to these whispers of a Moonless past. The prevailing theory, supported by a preponderance of evidence, posits that the Moon coalesced from the debris of a cataclysmic collision between the nascent Earth and a Mars-sized object early in our planet's history. This impact, a cosmic drama of unimaginable scale, is believed to have ejected a vast cloud of molten rock and dust into orbit, which gradually accreted to form our celestial companion. The Moon, thus born from the crucible of cosmic violence, has been a constant companion to Earth for billions of years, its gravitational embrace shaping our planet's tides, stabilizing its axial tilt, and influencing the very rhythm of life itself.

To imagine a world without the Moon is to envision a fundamentally different Earth, a planet whose history, environment, and perhaps even the course of its evolution would have been profoundly altered.

Despite the weight of scientific evidence, the allure of a Moonless past persists. The persistence of ancient myths and legends, the tantalizing hints in the geological record, and the inherent human fascination with the unknown all contribute to the enduring appeal of this enigmatic hypothesis. Perhaps, like the ruins of a forgotten civilization, these vestiges of a pre-lunar world offer glimpses into a distant past, a time before the Moon graced our skies, a time when the Earth danced alone in the celestial ballet, bathed in the pristine light of the stars. Or perhaps, they are merely echoes of a collective imagination, projections of our deepest fears and anxieties onto the canvas of the cosmos.

Whatever the truth may be, the contemplation of a Moonless Earth serves as a reminder of the vastness of time, the dynamism of our cosmic environment, and the enduring human quest to understand our place in the grand tapestry of the universe.

Chapter 5: The Moon's Eternal Vigil

In the celestial ballet of our solar system, the Earth and its Moon waltz in a graceful embrace, their fates intertwined since the dawn of time. While Earth, the verdant cradle of life, teems with vibrant diversity, the Moon, its silent companion, stands as a sentinel, a celestial shield guarding our planet from the slings and arrows of a perilous cosmos.

For countless millennia, the Moon has borne the brunt of cosmic bombardment, its cratered surface a testament to its unwavering defense. Meteors, asteroids, and comets, celestial projectiles hurtling through the void, have found their fiery demise upon the lunar surface, their destructive potential thwarted before they could reach Earth's fragile biosphere. The Moon's gravitational influence, a gentle yet potent force, acts as a celestial shepherd, nudging errant celestial bodies away from Earth's path. Like a watchful guardian, it deflects these cosmic threats, ensuring that their trajectories do not intersect with our own.

But the Moon's role as Earth's protector extends far beyond mere physical defense. It is a source of tidal forces, the rhythmic ebb and flow of our oceans, a vital factor in the delicate balance of Earth's climate and ecosystems. The Moon's gravitational pull also stabilizes Earth's rotational axis, preventing it from wobbling erratically, thus safeguarding the planet from extreme climatic shifts.

Moreover, the Moon serves as a celestial muse, inspiring poets, artists, and philosophers throughout history. Its luminous presence in the night sky has captivated humanity's imagination, igniting a sense of wonder and curiosity about the universe beyond our terrestrial confines.

As we venture further into the cosmos, the Moon remains a stepping stone, a celestial launchpad for our exploration of the solar system. The lessons learned from lunar missions have paved the way for our ambitious endeavors beyond Earth's cradle, propelling us towards the vast expanse of the cosmos.

The Moon is not merely a celestial body orbiting Earth; it is a protector, a stabilizer, a muse, and a stepping stone. Its unwavering presence has safeguarded our planet, nurtured life, and ignited our curiosity about the universe. As we continue to explore the cosmos, let us remember the Moon's invaluable role in our past, present, and future, and honor its eternal vigil, a testament to the profound connection between Earth and its celestial guardian.

The hypothetical cataclysm wrought by "Mohammed" serves as a potent *memento mori* for our seemingly secure terrestrial sphere. While the heavens may appear serene and immutable to the casual observer, a discerning eye perceives the lurking chaos, the ever-present potential for celestial bodies to stray from their appointed paths and wreak havoc upon the unsuspecting Earth. Though the likelihood of such a calamitous event remains, thankfully, remote, the consequences are of such magnitude that prudence dictates we devote ourselves to the noble task of planetary defense.

To this end, a comprehensive understanding of Near-Earth Objects (NEOs) is paramount. This celestial menagerie encompasses a diverse array of bodies, each with its own peculiar characteristics and potential for mischief. First, we must consider the asteroids, those rocky fragments, remnants of the solar system's primordial chaos. These wanderers, ranging in size from mere pebbles to behemoths dwarfing our terrestrial cities, typically reside within the asteroid belt, a vast expanse between Mars and Jupiter. Yet, the celestial dance is not without its occasional missteps. Gravitational perturbations, be they the subtle tug of Jupiter's immense gravity or the jarring collision with a fellow traveler, can send these asteroids careening towards our vulnerable world.

Then there are the comets, those icy apparitions that grace our skies with their ethereal tails. Hailing from the frigid depths of the outer solar system, these "dirty snowballs," as they are affectionately termed by astronomers, are propelled sunward by the inexorable forces of gravity. Their high velocities and often erratic trajectories render them particularly troublesome, for their arrival can be as sudden and unpredictable as a bolt from the blue.

Finally, we must not overlook the meteoroids, those lesser brethren of the asteroids and comets. Ranging in size from mere dust grains to formidable boulders, these celestial projectiles bombard our atmosphere daily, creating the fleeting spectacle of "shooting stars." While most are consumed in fiery immolation, the larger amongst them can penetrate our protective veil and strike the Earth, leaving scars upon the land and serving as a stark reminder of our precarious existence.

The threat posed by these celestial interlopers is not merely a figment of the imagination, a phantom menace conjured by overactive minds. The very ground beneath our feet bears witness to the destructive power of these cosmic projectiles. The Chicxulub impact, that cataclysmic event that extinguished the reign of the dinosaurs some sixty-six million years ago, serves as a chilling testament to the destructive potential of these celestial wanderers. More recently, the Tunguska event of 1908, which leveled a vast expanse of Siberian forest with the force of a thousand suns, reminds us that even in modern times, we are not immune to the capricious whims of the cosmos.

Recognizing this ever-present danger, the enlightened nations of the world have embarked upon a concerted effort to detect, track, and characterize these celestial interlopers. A global network of sentinels, both terrestrial and celestial, scans the heavens with tireless vigilance, seeking to identify potential threats and predict their trajectories. Ground-based telescopes, perched atop remote mountain peaks, peer into the depths of space, their powerful optics capturing the faintest glimmer of light reflected from these distant wanderers. The Pan-STARRS observatory in Hawaii, with its panoramic view of the cosmos, and the Catalina Sky Survey in Arizona, tirelessly scanning the heavens, stand as vanguards in this noble pursuit.

Complementing these terrestrial sentinels are their celestial counterparts, orbiting observatories that transcend the limitations of our atmosphere. NASA's NEOWISE telescope, for instance, with its infrared vision, pierces the cosmic veil, revealing objects hidden from the view of Earth-bound instruments.

The vast torrent of data generated by these tireless sentinels is then processed by sophisticated algorithms, mathematical engines that calculate orbits, predict future trajectories, and assess the likelihood of a calamitous encounter with our planet. The Minor Planet Center, a bastion of astronomical knowledge, serves as a clearinghouse for this vital information, disseminating timely alerts to the global community.

Yet, the mere identification of these celestial threats is but the first step in our quest for planetary security. We must also develop and deploy effective strategies to mitigate the risk of impact, to deflect these cosmic projectiles from their destructive paths. One promising approach is the kinetic impactor, a technique that employs a spacecraft as a celestial battering ram. By colliding with an NEO at high velocity, this intrepid emissary can impart sufficient momentum to alter its trajectory, nudging it away from a collision course with Earth.

Another intriguing concept is the gravity tractor, a celestial shepherd that gently guides an NEO away from danger through the subtle tug of its gravitational field. This method, though requiring a longer lead time, offers a more controlled and predictable means of deflection.

For those truly perilous situations, where time is of the essence, the nuclear option remains a possibility, albeit a controversial one. While the detonation of a nuclear device in the vicinity of an NEO could effectively deflect or disrupt it, the potential for unintended consequences, such as fragmentation and the creation of radioactive debris, must be carefully weighed.

And should all our efforts prove futile, and a collision become unavoidable, we must be prepared to implement civil defens measures. Evacuation plans, shelters, and disaster preparednes protocols will be crucial in minimizing casualties and mitigatin; the inevitable chaos.

The development and implementation of these planetary defense strategies necessitate a global effort, a concerted endeavor that transcends national borders and politica ideologies. The hypothetical impact of "Mohammed" serves as a potent reminder of our shared vulnerability and underscores the imperative for international cooperation in safeguarding our fragile world.

In the subsequent section, we shall return to the hypothetical impact of "Mohammed" and examine its specific implications for human activities in space, exploring the challenges and opportunities that this celestial cataclysm presents for the future of lunar exploration.

Having surveyed the diverse array of celestial bodies that constitute the NEO population, and having acknowledged the very real threat they pose to our terrestrial abode, let us now turn our attention to the specific challenges and opportunities presented by the hypothetical impact of "Mohammed" on the Moon. This celestial cataclysm, though confined to our lunar neighbour, would have far-reaching implications for human activities in space, forcing us to re-evaluate our strategies for lunar exploration and colonization.

The impact of "Mohammed" would irrevocably alter the lunar landscape, creating a vast crater that would dominate the Moon's southern hemisphere. This newly formed crater, with its exposed layers of lunar crust and mantle, would become a focal point for scientific investigation, offering a unique opportunity to study the Moon's internal structure and geological history.

However, the impact would also pose challenges for future lunar missions. The vast debris field surrounding the crater could pose a hazard to spacecraft, while the altered gravitational field could affect landing trajectories and orbital stability. The increased meteoroid activity in the Earth-Moon system could also necessitate enhanced shielding for spacecraft and lunar habitats.

The impact of "Mohammed" would force us to adapt our strategies for lunar exploration and colonization. Future missions would need to account for the changed lunar environment, incorporating new technologies and operational procedures to ensure the safety and efficiency of lunar operations.

The impact could also stimulate innovation, driving the development of new technologies for lunar resource utilization, habitat construction, and planetary defense. The need to adapt to a more dynamic and hazardous lunar environment could spur advancements in robotics, artificial intelligence, and space-based infrastructure.

Despite the challenges posed by the impact of "Mohammed," the Moon would remain a compelling destination for human exploration and scientific discovery. The altered lunar landscape, with its newly exposed geological formations and potential for resource extraction, would offer new opportunities for scientific research and technological development.

The impact could also serve as a catalyst for international cooperation, as nations pool their resources and expertise to overcome the challenges and capitalize on the opportunities presented by the new lunar environment. The shared pursuit of lunar exploration could foster a sense of global unity and inspire a new generation of scientists, engineers, and explorers.

The hypothetical impact of "Mohammed" on the Moon presents a unique set of challenges and opportunities for human civilization. It is a scenario that forces us to confront the dynamic nature of our cosmic environment and to adapt our strategies for space exploration and colonization.

However, it is also a scenario that highlights the resilience and ingenuity of humankind. Faced with a changed lunar landscape, we would undoubtedly rise to the challenge, developing new technologies, forging new partnerships, and pushing the boundaries of human knowledge and exploration.

The Moon, though scarred, would remain a beacon in our night sky, a symbol of our enduring fascination with the universe and our unwavering determination to explore its mysteries. The impact of "Mohammed" would not mark the end of our lunar ambitions, but rather a new chapter in the ongoing saga of human exploration and discovery.

In the concluding chapter, we shall reflect on the broader implications of our exploration, considering the lessons learned from the hypothetical impact of "Mohammed" and contemplating the future of humanity in a cosmos teeming with both peril and promise.

Chapter 6: Reflections on a Celestial Cataclysm: Humanity's Future Amongst the Stars

The hypothetical impact of "Mohammed" upon the Moon, a celestial drama played out upon the stage of our imagination, has served as a lens through which to examine humanity's relationship with the cosmos. This speculative exercise, grounded in the principles of science yet imbued with the spirit of philosophical inquiry, has led us on a journey through the annals of history, the depths of space, and the intricacies of human society.

We have witnessed the Moon's profound influence upon human culture, serving as a source of wonder, inspiration, and even religious devotion. We have delved into the mechanics of celestial impacts, exploring the destructive potential of these cosmic events and the challenges they pose to our terrestrial security. And we have contemplated the resilience of human civilization, our capacity to adapt, innovate, and ultimately transcend even the most daunting of challenges.

The hypothetical impact of "Mohammed," though a fictional event, serves as a potent reminder of our precarious existence within the vast expanse of the universe. It underscores the dynamic nature of our cosmic environment, the ever-present potential for celestial events to disrupt the delicate balance of our world. Yet, it also highlights the indomitable spirit of humankind, our capacity to confront adversity with courage and ingenuity. Faced with a changed lunar landscape, we would undoubtedly rise to the challenge, harnessing our scientific knowledge, technological prowess, and collaborative spirit to forge a new path amongst the stars.

The lessons learned from this hypothetical cataclysm extend far beyond the realm of lunar exploration. They speak to the very essence of human existence, our place within the grand tapestry of the cosmos, and our responsibility to safeguard our planet and its inhabitants.

The impact of "Mohammed," though a fictional event, serves as a call to action, urging us to invest in scientific research, develop planetary defense strategies, and foster international cooperation in the face of shared cosmic threats. It is a call to embrace our role as stewards of this planet, to protect its fragile ecosystems, and to ensure the survival and prosperity of future generations. And it is a call to continue our exploration of the cosmos, to push the boundaries of human knowledge and understanding, and to seek our destiny amongst the stars.

For in the words of the ancient Roman poet, Lucretius, "Nothing is ever born from nothing." The universe, in all its vastness and complexity, is a product of natural laws, a symphony of matter and energy governed by the inexorable forces of gravity and motion. The hypothetical impact of "Mohammed" is but one note in this cosmic symphony, a fleeting moment in the grand sweep of time. Yet, it serves as a powerful reminder of our interconnectedness with the universe, our dependence upon its delicate balance, and our responsibility to preserve its wonders for generations to come.

As we stand upon the threshold of a new era of space exploration, poised to venture beyond the confines of our terrestrial cradle, let us carry with us the lessons learned from this hypothetical cataclysm. Let us approach the cosmos with humility and respect, mindful of its power and beauty, yet ever determined to unlock its secrets and fulfill our destiny amongst the stars. For the future of humanity lies not solely upon this Earth, but amongst the infinite expanse of the cosmos, where new worlds await our discovery, new challenges beckon our ingenuity, and new wonders inspire our dreams.

The hypothetical impact of "Mohammed" on the Moon presents a unique set of challenges and opportunities for human activities in space, particularly for the future of lunar exploration and the development of a permanent lunar base.

The impact would pose a significant threat to any existing or planned lunar infrastructure. The massive crater and the extensive debris field would create hazardous conditions for lunar missions, requiring careful planning and advanced navigation systems to ensure the safety of astronauts and equipment. The dust and debris ejected by the impact could also damage sensitive instruments and solar panels, disrupting power supplies and communication networks. The potential for increased meteoroid activity in the years following the impact would further complicate lunar operations, requiring robust shielding and contingency plans to protect lunar habitats and equipment.

The impact would fundamentally alter the lunar environment, requiring adjustments to existing plans for lunar exploration and colonization. The altered gravitational field, the presence of new geological features, and the potential for increased seismi activity would necessitate a re-evaluation of landing sites, habita designs, and resource extraction strategies.

The impact could also reveal new scientific opportunities. The exposed layers of lunar crust and mantle in the impact crate would provide valuable insights into the Moon's internal structure and geological history. The study of the debris field could shed light on the composition and origin of "Mohammed" and contribute to our understanding of asteroid impacts.

The hypothetical impact of "Mohammed" underscores the importance of developing effective planetary defense strategies, not only for Earth but also for our celestial neighbors. The Moon, lacking a substantial atmosphere, is particularly vulnerable to impacts, and any future lunar base would need to be equipped with robust defenses against meteoroid strikes. The development of early warning systems, deflection technologies, and protective shielding would be crucial in ensuring the safety and sustainability of lunar settlements. The experience gained from mitigating the impact of "Mohammed" could provide valuable lessons for protecting future lunar outposts and ensuring the long-term viability of human activities on the Moon.

Despite the challenges, the impact of "Mohammed" could also usher in a new era of lunar exploration. The altered lunar landscape, with its new geological features and potential for resource extraction, could attract renewed interest from scientists, entrepreneurs, and space agencies. The impact could also stimulate innovation in lunar transportation, habitat design, and resource utilization. The need to adapt to a changed lunar environment could lead to the development of new technologies and strategies for lunar exploration, paving the way for a more sustainable and resilient human presence on the Moon.

The Moon, even in its altered state, remains a crucial stepping stone for humanity's further exploration of the solar system. Its proximity to Earth, its abundant resources, and its scientific potential make it an ideal location for developing the technologies and infrastructure needed for future missions to Mars and beyond. The hypothetical impact of "Mohammed" serves as a reminder of the challenges and opportunities that await us as we venture further into the cosmos. It is a call to action, urging us to develop the knowledge, technology, and international cooperation needed to safeguard our planet and explore the vast expanse of the universe.

While the impact of "Mohammed" is a hypothetical event, it serves as a valuable case study for examining the challenges and opportunities associated with planetary defense. By analyzing the potential consequences of this impact, we can gain insights into the strategies and technologies needed to protect Earth from future asteroid threats.

The first line of defense against NEO impacts is early detection and characterization. Telescope surveys, both ground-based and space-based, play a crucial role in identifying and tracking potentially hazardous objects. The sooner an NEO is detected, the more time we have to assess its trajectory and develop mitigation strategies. In the case of "Mohammed," early detection would be crucial in determining the size, composition, and trajectory of the asteroid. This information would allow scientists to accurately predict the impact location and assess the potential consequences for Earth and the Moon.

Once a potentially hazardous NEO is identified, various deflection strategies can be employed to alter its trajectory and prevent an impact. The choice of strategy depends on factors such as the size and mass of the NEO, the lead time available for intervention, and the technological capabilities at our disposal.

Kinetic Impactor: This technique involves using a spacecraft to collide with an NEO, transferring momentum and altering its course. The effectiveness of this method depends on the mass of the impactor and the velocity of the collision. In the case of "Mohammed," a kinetic impactor could potentially be used to deflect the asteroid away from the Moon, preventing the impact altogether.

Gravity Tractor: This concept involves using a spacecraft to exert a gravitational pull on an NEO, gradually altering its trajectory over time. This method is particularly suitable for smaller NEOs and requires a longer lead time for intervention. While it might not be feasible to completely deflect "Mohammed" with a gravity tractor, it could potentially be used to nudge it onto a slightly different trajectory, reducing the severity of the impact.

Nuclear Option: The use of nuclear weapons to deflect or disrupt an NEO remains a controversial option. While it could be effective in certain scenarios, it also carries significant risks, including the potential for fragmentation and the creation of radioactive debris. In the case of "Mohammed," the nuclear option might be considered as a last resort if other deflection methods prove infeasible.

Planetary defense is a global challenge that requires international cooperation and coordination. No single nation possesses the resources and expertise to effectively address the threat of NEO impacts. Sharing information, coordinating observation efforts, and developing joint mitigation strategies are essential for protecting our planet. The hypothetical impact of "Mohammed" highlights the need for a robust international framework for planetary defense. This framework should include protocols for sharing NEO data, coordinating deflection efforts, and establishing clear lines of responsibility in the event of an impending impact.

Public awareness and education are crucial components of planetary defense. Informing the public about the risks posed by NEOs, the strategies being developed to mitigate those risks, and the importance of international cooperation can help build support for planetary defense initiatives and ensure a coordinated response in the event of an impact threat. The hypothetical impact of "Mohammed" could serve as a valuable opportunity to raise public awareness about planetary defense. By communicating the potential consequences of such an event, we can engage the public in discussions about the importance of investing in NEO research, developing deflection technologies, and fostering international collaboration.

The hypothetical impact of "Mohammed" on the Moon is a wake-up call, reminding us of the dynamic nature of our cosmic environment and the potential for catastrophic events beyond our control. However, it is also a call to action, urging us to develop the knowledge, technology, and international cooperation needed to safeguard our planet and ensure the long-term survival of human civilization. By investing in NEO research, developing deflection technologies, and fostering international collaboration, we can become the sentinels of our own sky, protecting our planet from the potential threats lurking in the vastness of space.

The hypothetical impact of "Mohammed" serves as a reminder that our destiny is not predetermined, but rather shaped by our choices and our actions. By embracing the challenge of planetary defense, we can ensure that humanity continues to thrive on Earth and beyond, even in the face of cosmic adversity.

In the concluding section of this chapter, we will reflect on the broader philosophical and ethical implications of planetary defense, considering the responsibility that comes with the power to alter the course of celestial objects and safeguard the future of our planet.

This shift in perspective could lead to a renewed appreciation for the fragility of our planet and the interconnectedness of life on Earth with the cosmos. It could also inspire a deeper sense of humility, recognizing our place within the vast expanse of the universe and the limitations of our control over natural forces.

Despite the profound challenges posed by a Moonless world, humanity has demonstrated throughout history a remarkable capacity for resilience and adaptation. We have faced numerous natural disasters, wars, and pandemics, each time emerging stronger and more resourceful. The loss of the Moon, while unprecedented, would likely trigger a similar response, prompting innovation, cooperation, and a re-evaluation of our priorities.

Scientists and engineers would undoubtedly rise to the challenge, developing new technologies to adapt to the altered environment, mitigate the risks posed by the fragmented Moon, and explore the new scientific opportunities presented by the transformed lunar landscape. Artists and writers would find new ways to express the human experience in a Moonless world, exploring themes of loss, resilience, and the enduring human spirit.

The hypothetical impact of "Mohammed" on the Moon represents a potential turning point in human history. It is a scenario that challenges our assumptions about the stability of our cosmic environment and forces us to confront the potential for catastrophic events beyond our control.

However, it is also a scenario that highlights the resilience and adaptability of human civilization. The loss of the Moon, while profoundly disruptive, could also serve as a catalyst for innovation, cooperation, and a renewed appreciation for our interconnectedness with the cosmos. It could inspire us to explore new frontiers, develop sustainable technologies, and build a more resilient and adaptable civilization.

The Moon, even in its altered state, would remain a powerful symbol in our collective consciousness, a reminder of our enduring fascination with the universe and our unwavering determination to overcome adversity. The impact of "Mohammed" would not be the end of our story, but rather a new chapter in the ongoing saga of human exploration, adaptation, and our quest to understand our place in the cosmos. In the following chapter, we will delve into the potential long-term implications of a Moonless world, exploring the possible evolutionary paths of human civilization and our relationship with the cosmos in the absence of our celestial companion.

The hypothetical destruction of the Moon, while a cataclysmic event, would not mark the end of human history. Instead, it would usher in a new era, one defined by profound environmental shifts, technological adaptation, and a fundamental reorientation of humanity's relationship with the cosmos.

The most immediate and pervasive consequence of a Moonless world would be the disruption of Earth's tides. The Moon's gravitational pull is the primary driver of our ocean's rhythmic ebb and flow, shaping coastal ecosystems, influencing marine life cycles, and even playing a subtle role in our planet's climate. Without the Moon, tides would be significantly reduced, becoming mere whispers of their former selves, driven primarily by the weaker gravitational influence of the Sun.

This diminished tidal force would have far-reaching consequences. Coastal ecosystems, adapted to the regular inundation and exposure of tidal zones, would be thrown into disarray. Marine life, dependent on tidal currents for nutrient transport and migration patterns, would face new challenges. Human settlements and infrastructure along coastlines, built with the expectation of predictable tides, would become vulnerable to erosion and flooding.

The loss of the Moon's stabilizing influence on Earth's axial tilt could also lead to more extreme and unpredictable climate variations over long timescales. While the exact nature of these changes is difficult to predict, they could result in more frequent ice ages, dramatic shifts in weather patterns, and significant challenges for agriculture and human settlements.

The absence of the Moon would transform the nocturnal landscape. The night sky, once illuminated by its gentle radiance, would become significantly darker, revealing the full splendor of the stars and the Milky Way. This newfound celestial clarity could inspire a resurgence of astronomical observation and a deeper appreciation for the vastness of the universe.

However, the increased darkness could also have practical implications. Nocturnal animals, adapted to hunting and navigating under the Moon's light, would face new challenges. Human activities that rely on nighttime illumination, such as transportation and agriculture, would require adaptation and innovation.

The loss of the Moon would spur a wave of technological innovation as humanity seeks to adapt to the altered environment and mitigate the challenges posed by its absence. Engineers and scientists would develop new methods for predicting and managing coastal erosion, harnessing alternative energy sources to compensate for the loss of tidal power, and creating artificial lighting systems to illuminate the night.

Space exploration would likely take on a renewed urgency. With the Moon no longer a viable destination for colonization or resource extraction, humanity might accelerate its efforts to reach Mars and other celestial bodies, seeking new frontiers for exploration and settlement.

The disappearance of the Moon could trigger a profound philosophical and spiritual reawakening. The loss of this celestial icon, a constant presence throughout human history, could prompt a re-evaluation of our place in the cosmos, our relationship with the natural world, and the meaning of existence itself. Philosophers and theologians would grapple with new questions about the nature of reality, the fragility of existence, and the role of humanity in a universe devoid of its familiar celestial companion. Artists and writers would explore these themes in new and creative ways, expressing the anxieties, hopes, and aspirations of a civilization navigating a transformed world.

Despite the profound challenges and uncertainties of a Moonless world, the human spirit would undoubtedly endure. Throughout history, we have faced adversity with resilience, ingenuity, and a determination to overcome obstacles. The loss of the Moon, while a significant loss, would not extinguish the flame of human curiosity, creativity, and the pursuit of knowledge. We would adapt, innovate, and find new ways to thrive in a world without its familiar celestial companion. We would explore new frontiers, unlock new mysteries, and continue our journey of discovery, carrying with us the legacy of the Moon and the lessons learned from its absence.

In the concluding chapter of this work, we will reflect on the broader implications of the "Mohammed" scenario, considering its potential to reshape our understanding of the cosmos, our place within it, and the future of human civilization.

Yet, amidst the lamentations and the ashes, the indomitable spirit of man would rise from the celestial ashes, much like a phoenix from the flames. The catastrophe, though a cataclysm, would also serve as a crucible, testing the mettle of our civilization and revealing the true strength of our resolve. For even in the face of cosmic ruin, humanity, armed with its ingenuity and resilience, would chart a new course, much like a ship captain adjusting his sails to an unexpected squall.

Chapter 8: Homo Lunaris: Humanity in a Moonless Future

The hypothetical loss of the Moon, a celestial companion that has graced our skies for billions of years, would undoubtedly mark a profound turning point in human history. It would challenge our understanding of the cosmos, reshape our relationship with the natural world, and potentially alter the very trajectory of our civilization. This chapter explores the long-term implications of a Moonless world, speculating on the potential evolutionary paths of humanity and our adaptation to a fundamentally altered celestial landscape.

The savants of this epoch, those inheritors of Galileo's inquisitive spirit and Newton's perspicacious mind, would cast their gaze upon the fractured Moon, their instruments trained upon the celestial ruins. They would measure, analyze, and theorize, seeking to unravel the mysteries of this cosmic cataclysm, much like a detective reconstructing a crime from the fragments of evidence. Their relentless pursuit of knowledge, a beacon amidst the darkness, would illuminate the path forward, guiding humanity towards a deeper understanding of its place in the vast cosmic tapestry.

The absence of the Moon would necessitate a profound reorientation of human existence. The familiar rhythms of life, governed by the lunar cycle and its influence on tides and light patterns, would be disrupted, requiring adaptation and innovation. This process of adaptation could lead to the emergence of a new kind of human, Homo lunaris, characterized by a unique set of physical, psychological, and cultural traits.

Physiological Adaptations: Without the Moon's gravitational influence, the human body might undergo subtle physiological changes over generations. Bone density, muscle mass, and even circadian rhythms could be affected, leading to a new human physique better suited to a world with diminished tidal forces and altered light cycles.

Technological Mastery: The loss of the Moon would spur a new wave of technological innovation. Humanity would develop advanced systems for artificial illumination, climate control, and coastal management to mitigate the challenges of a Moonless world. Space exploration would likely accelerate, with a focus on establishing settlements on Mars and other celestial bodies, securing humanity's future beyond Earth.

Psychological Resilience: The psychological impact of losing the Moon would be profound, but humanity would ultimately adapt. New cultural narratives, philosophical perspectives, and spiritual practices would emerge, helping us to make sense of the altered cosmic landscape and find meaning in a world without its familiar celestial companion.

The absence of the Moon could also lead to a heightened awareness of our place in the cosmos. With the night sky unveiled in its full splendor, the stars and the Milky Way would become more prominent, reminding us of the vastness and complexity of the universe. This could foster a deeper appreciation for the interconnectedness of all things and a greater sense of responsibility for our planet and its future.

The loss of the Moon could also inspire a renewed focus on scientific exploration and a deeper understanding of the cosmos. With our celestial neighbor gone, we might turn our attention to the mysteries of the universe with greater urgency, seeking answers to fundamental questions about our origins, our place in the cosmos, and the potential for life beyond Earth.

The absence of the Moon could lead to divergent evolutionary paths for human civilization. Some societies might embrace technological solutions, creating artificial environments that mimic the Moon's influence on tides and light cycles. Others might adapt to the new reality, developing sustainable practices that harmonize with the natural rhythms of a Moonless world. This divergence could lead to the emergence of distinct cultures and civilizations, each with its own unique relationship with the cosmos and its own vision for the future of humanity. The loss of the Moon, while a shared experience, could ultimately lead to a greater diversity of human expression and a richer tapestry of cultural narratives.

The hypothetical "Mohammed" scenario challenges us to confront fundamental questions about the nature of existence, the role of humanity in the cosmos, and the meaning of life in a universe devoid of its familiar celestial companion. It is a scenario that invites speculation, contemplation, and a renewed appreciation for the fragility and resilience of life on Earth.

While the loss of the Moon would undoubtedly be a profound loss, it would not diminish the human spirit. We would adapt, innovate, and continue our quest for knowledge and meaning, carrying with us the legacy of the Moon and the lessons learned from its absence. The Moon, even in its absence, would remain a powerful symbol in our collective consciousness, a reminder of our enduring fascination with the universe and our unwavering determination to overcome adversity. The hypothetical impact of "Mohammed" would not mark the end of our story, but rather a new chapter in the ongoing saga of human exploration, adaptation, and our quest to understand our place in the cosmos. While the sober mind of the natural philosopher may recoil at the notion of celestial bodies obedient to the whims of deities, a survey of the world's mythologies reveals a persistent human tendency to ascribe agency and intention to the cosmic drama. From the thunderbolts of Jupiter to the celestial chariot of the Sun God Ra, humanity has long populated the heavens with a panoply of supernatural beings, their actions shaping the destinies of mortals and the very fabric of the universe.

In the annals of Islam, a faith born amidst the sands of Arabia and destined to spread its dominion across vast swathes of the globe, we find a particularly striking example of such celestial intervention: the legendary splitting of the Moon. Attributed to the Prophet Mohammed, peace be upon him, this miraculous event, as recounted in the sacred texts of the Quran and the venerable traditions of the Hadith, serves as a potent symbol of divine power and a testament to the Prophet's privileged connection to the Almighty.

Yet, the Islamic tradition is but one thread in the rich tapestry of world mythology. Across continents and cultures, from the snow-capped peaks of the Himalayas to the sun-drenched shores of the Mediterranean, humanity has woven tales of celestial intervention, imbuing the cosmos with a vibrant cast of gods, demons, and heroes, their actions shaping the destinies of mortals and the very fabric of the universe.

In the venerable land of Bharat, where the ancient wisdom of the Vedas and the Puranas has been passed down through countless generations, we encounter a pantheon of deities whose powers extend to the celestial realm. The churning of the cosmic ocean, a feat of unimaginable scale undertaken by gods and demons alike, speaks to the Hindu conception of a universe in constant flux, where even the celestial spheres are subject to the whims of divine beings.

Lord Krishna, that beloved avatar of Vishnu, the preserver of the universe, is said to have lifted Mount Govardhan with his divine hand, shielding the villagers of Vrindavan from the wrath of Indra, the king of the gods. This act of cosmic defiance, a testament to Krishna's boundless compassion and power, echoes the theme of divine intervention in the natural order, a theme that resonates across cultures and epochs.

The descent of the Ganges, that sacred river revered by millions, is another tale of celestial intervention in Hindu mythology. As the story goes, the mighty river, originating in the heavens, threatened to overwhelm the Earth with its torrential flow. But Lord Shiva, the ascetic god of destruction and renewal, intervened, receiving the river upon his matted locks and gently releasing it onto the Earth, transforming a potential catastrophe into a source of life and spiritual sustenance.

Even the seemingly predictable phenomena of eclipses are imbued with mythological significance in the Hindu tradition. Rahu, a demonic entity, is said to devour the Sun or Moon, causing their temporary disappearance from the sky. This celestial drama, a battle between light and darkness, good and evil, serves as a reminder of the constant struggle between opposing forces in the universe and the precarious balance that governs the cosmos.

Beyond the shores of India, in the lands of ancient Greece and Rome, where philosophy and mythology intertwined, we find a pantheon of gods and goddesses whose powers extended to the celestial realm. Apollo, the radiant god of light and music, drove his chariot across the sky, bringing the Sun's life-giving rays to the world. Diana, the huntress goddess, was associated with the Moon, her silvery arrows piercing the darkness of night.

In the Norse sagas, where tales of valor and adventure unfold amidst a backdrop of icy landscapes and stormy seas, the gods Thor and Odin wielded their celestial powers to shape the destinies of mortals and maintain the cosmic order. Thor, with his mighty hammer Mjolnir, battled giants and monsters, his thunderous blows echoing across the heavens. Odin, the wise and enigmatic Allfather, presided over the realm of the gods, his watchful eye ever vigilant against the forces of chaos.

These narratives, drawn from diverse cultures and traditions, demonstrate the enduring human fascination with the cosmos and our attempts to understand our place within the grand scheme of the universe. They reflect our desire to imbue the celestial sphere with meaning and purpose, to populate it with beings that embody our hopes, fears, and aspirations.

While the scientific mind may seek to explain these celestial phenomena through the lens of natural laws and physical processes, the power of myth and storytelling endures. These narratives continue to shape our understanding of the cosmos, our place within it, and the meaning of human existence.

The hypothetical splitting of the Moon by a meteor, while grounded in scientific principles, also resonates with these ancient myths and legends. It is a story that speaks to our fascination with the cosmos, our vulnerability to natural forces, and our enduring quest for meaning in a universe full of wonder and uncertainty.

In the next chapter, we shall return to the hypothetical impact of "Mohammed" and explore its implications for the future of human civilization, considering the challenges and opportunities presented by a world forever altered by the loss of its celestial companion.

The absence of the Moon would not merely be a physical phenomenon; it would be a cultural and psychological earthquake, shaking the foundations of human understanding and forcing a re-evaluation of our place in the cosmos. The night sky, once dominated by the Moon's gentle glow, would be transformed into a canvas of unfathomable depth, studded with a million glittering stars, the Milky Way stretching across the heavens like a luminous river.

This newfound celestial grandeur could inspire a resurgence of astronomical curiosity, a renewed sense of wonder at the vastness of the universe, and a deeper appreciation for the interconnectedness of all things. Artists, writers, and musicians would undoubtedly draw inspiration from this altered celestial landscape, expressing the anxieties, hopes, and aspirations of a civilization navigating a transformed world. New myths and legends would emerge, weaving narratives around the lost Moon, the celestial cataclysm that reshaped our world, and the resilience of the human spirit in the face of cosmic adversity.

The loss of the Moon could also lead to a re-evaluation of our relationship with time. The lunar cycle, with its predictable phases, has long served as a natural clock, shaping our calendars, agricultural practices, and even our biological rhythms. Without the Moon's rhythmic presence, we might develop new ways of measuring time, perhaps based on the cycles of the Sun, the seasons, or even the rhythms of our own bodies.

The profound psychological impact of losing the Moon could trigger a shift in human consciousness. The familiar sense of comfort and stability provided by the Moon's presence would be replaced by a heightened awareness of our vulnerability and our dependence on the delicate balance of our planet's ecosystem. This could lead to a deeper appreciation for the interconnectedness of all life and a greater sense of responsibility for safeguarding the future of our planet.

The absence of the Moon could also foster a more profound sense of existential awareness. The night sky, unveiled in its full splendor, would serve as a constant reminder of the vastness and mystery of the universe, prompting questions about our origins, our purpose, and our place in the grand scheme of existence. This could lead to a resurgence of philosophical and spiritual inquiry, as we seek to make sense of our existence in a universe devoid of its familiar celestial companion.

The challenges posed by a Moonless world would undoubtedly spur a wave of technological innovation. Humanity would need to develop new methods for managing coastal erosion, harnessing alternative energy sources, and creating artificial lighting systems to illuminate the night. Space exploration would likely take on a renewed urgency, with a focus on establishing settlements on Mars and other celestial bodies, securing humanity's future beyond Earth.

The development of advanced technologies could also lead to new forms of human augmentation and enhancement. Genetic engineering, artificial intelligence, and nanotechnology could be employed to enhance our physical and cognitive capabilities, allowing us to adapt to the challenges of a Moonless world and explore the cosmos with greater resilience and efficiency.

The loss of the Moon could also necessitate a re-evaluation of our social and political structures. The challenges of adapting to a changed world, managing resources, and ensuring the survival of our species could lead to new forms of global cooperation and governance. The need for collective action and shared responsibility could transcend national boundaries, fostering a sense of global citizenship and a shared commitment to the well-being of humanity and the planet.

The hypothetical "Mohammed" scenario offers a glimpse into a future that is both challenging and full of possibilities. The loss of the Moon would undoubtedly be a profound loss, but it would not mark the end of human history. Instead, it would usher in a new era, one defined by adaptation, innovation, and a renewed appreciation for the fragility and resilience of life on Earth.

The future of humanity in a Moonless world is uncertain, but it is also full of potential. We may evolve into a new kind of human, Homo lunaris, adapted to the challenges of a transformed world and equipped with the technology and wisdom to navigate the complexities of the cosmos. We may forge new social structures, embrace new forms of consciousness, and continue our quest for knowledge and meaning in a universe full of wonder and uncertainty.

The loss of the Moon would be a profound loss, but it would not diminish the human spirit. We would adapt, innovate, and continue our journey of discovery, carrying with us the legacy of the Moon and the lessons learned from its absence. The hypothetical impact of "Mohammed" would not mark the end of our story, but rather a new chapter in the ongoing saga of human exploration, adaptation, and our quest to understand our place in the cosmos.

Bibliography

Mythology and Classical Sources

- Aristotle. *On the Heavens*. (Various editions)
- Apollonius of Rhodes. *Argonautica*. (Various editions)
- Plutarch. *Moralia*. (Various editions)
- Hesiod. *Theogony*. (Various editions)
- Homer. *Iliad* and *Odyssey*. (Various editions)
- Ovid. *Metamorphoses*. (Various editions)
- Virgil. *Aeneid*. (Various editions)
- The Bible. (Various editions)
- The Quran. (Various editions)
- Hadith. (Various collections)
- Mahabharata. (Various editions)
- Ramayana. (Various editions)

Scientific Sources

- Lewis, John S. *Rain of Iron and Ice: The Very Real Threat of Comet and Asteroid Bombardment*. Addison-Wesley, 1996.
- Morrison, David. *Planetary Defense: Protecting Earth from Asteroids*. Cambridge University Press, 2019.
- Neal, Valerie. *The History of the Moon*. Firefly Books, 2009.

This bibliography includes a mix of primary sources (e.g., The Quran, Hadith, classical works) and secondary sources (e.g., scientific books on asteroid impacts and lunar history) to provide a comprehensive overview of the topics explored in the text.

Don't miss out!

Visit the website below and you can sign up to receive emails whenever A.N.F. Simões publishes a new book. There's no charge and no obligation.

https://books2read.com/r/B-A-GEROC-GFTIF

BOOKS 2 READ

Connecting independent readers to independent writers.

Did you love *The Splitting of the Moon*? Then you should read *The Time Weaver's Tapestry*[1] by A.N.F. Simões!

[2]

This work embarks on a grand survey of history, archaeology, and modern physics, re-examining the past through the lens of quantum mechanics and the multiverse. It challenges the traditional linear view of time, exploring anomalies in the archaeological record and inconsistencies in historical narratives that suggest a more complex, multi-faceted past.

The work delves into the quantum revolution, the dethronement of time as an absolute entity, and the implications of a multiverse of histories, questioning the very nature of

---

1. https://books2read.com/u/4AlQ7N

2. https://books2read.com/u/4AlQ7N

historical "fact." It proposes a new model of historical inquiry, one that embraces uncertainty and the multiplicity of possible pasts, drawing on quantum concepts like superposition, entanglement, and non-locality to reinterpret the past.

This exploration extends to the realm of archaeology, proposing a "quantum archaeology" that seeks information encoded not just in physical artifacts but in the fabric of spacetime itself. It examines enigmatic artifacts, unexplained phenomena, and cryptic accounts that defy the limitations of their eras, suggesting a past more complex and enigmatic than previously imagined.

The work concludes by emphasizing the ethical implications of this new perspective on history and archaeology, calling for a responsible and compassionate approach to the exploration of the past and the shaping of the future.

Also by A.N.F. Simões

In the Shadow of the Kurgan: A Compendium of the
Indigenous People of Europe
The Cornfield
The Secret Rout
The Time Weaver's Tapestry
The New Republic of Manhood: Building a Better Future in a
World Gone Wrong
The Splitting of the Moon

About the Publisher

OLYSSIPO Publishing is an independent publishing house based in Sydney Australia. It was founded in 2023 with a mission to promote independent literature and give voice to underrepresented perspectives. OLYSSIPO is committed to publishing high-quality works of fiction, nonfiction, and poetry that challenge conventional thinking and explore the complexities of the human experience. OLYSSIPO is a strong advocate for freedom of expression and believes in the power of literature to foster dialogue and understanding. The name "OLYSSIPO" is the ancient Roman name for Lisbon, evoking a rich history of cultural exchange and intellectual curiosity.
A.N.F.Simoes ©2024